KU-511-512

THE UNIVERSE

Written by **Ian Graham**

TWO-CAN

First published in Great Britain in 1991 by
Two-Can Publishing Ltd
346 Old Street
London EC1V 9NQ

Copyright © Two-Can Publishing Ltd, 1991

Text copyright © Ian Graham 1991
Editorial and Design by Lionheart Books, London
Editor: Denny Robson   Picture Research: Jennie Karrach
Media conversion: Peter MacDonald, Una Macnamara and Vanessa Hersey
Studio Artwork: Radius

Printed and bound in Hong Kong

2 4 6 8 10 9 7 5 3

All rights reserved. No part of this publication may be reproduced, stored
in a retrieval system, or transmitted in any form or by any means,
electronic, mechanical, photocopying or otherwise, without prior
written permission of the copyright owner.

The JUMP! logo and the word JUMP! are registered trade marks.

British Library Cataloguing in Publication Data

Graham, Ian
    The Universe
    1. Universe
    1. Title
    523.1

ISBN 1-85434-098-0

Photographic Credits:
p.4-5 US Naval Observatory. p.6 Julian Baum. p.7 NASA. p.8 (top right) NASA. p.8 (bottom left) Julian Baum. p.9 NASA. p.10 US Naval Observatory. p.11 European
Southern Observatory. p.12 Royal Observatory, Edinburgh. p.13 Julian Baum. p.14 US Naval Observatory. p.15 NASA. p.16 Julian Baum. p.17 US Naval Observatory.
p.18 NASA. p.18-19 Julian Baum. p.20-21 Julian Baum. p.22 NASA. p.30,31 US Naval Observatory. Cover photo Art Directors Photo Library.

Illustration Credits:
All illustrations by Chris Forsey and Peter Bull except those on pages 24-28, which are by Graham Humphreys of Virgil Pomfret Artists.

CONTENTS

All words marked in **bold** can be found in the glossary

WHAT IS THE UNIVERSE?

The Universe contains everything that exists, from the smallest particles of matter, **atoms**, to the biggest star system. We live on the planet Earth, which circles, or **orbits**, a star called the Sun. Eight other planets also orbit the Sun. Most have **moons** orbiting around them. This collection of planets, moons and the Sun is called the Solar System. As far as we know, the Earth is the only place in the Universe where living creatures exist.

▲ The Solar System (1) is almost 12,000 million km (7,500 million miles) across. Beyond it, distances are so great that they are measured in **light-years**, the distance light travels in a year. Our galaxy (2) is 100,000 light-years across.

The Solar System seems to be very big, but it is only a tiny speck in space compared to the star system, or galaxy, that it belongs to. Our galaxy, the Milky Way, is just one of perhaps 100,000 million galaxies in the Universe. It is impossible to imagine how big the Universe is. Even if we could travel as fast as a beam of light, it would take 180,000 years to reach another galaxy. Because of the great distances between them, galaxies are sometimes called island Universes.

◄ The Pleiades are a group of nearly 400 stars, 400 light-years away from Earth. Seven of them, the Seven Sisters, can be seen in the night sky without a telescope.

▲ The Milky Way belongs to a group of over 30 galaxies called the Local Group (3). Beyond the Local Group, there are galaxies as far away as telescopes can see (4). The furthest seen so far is almost 14,000 million light-years away.

START OF THE UNIVERSE

There have been many theories about how and when the Universe began. Scientists now believe the Universe began 20,000 million years ago. At that time there were no stars, galaxies or planets. One theory now believed is that everything in the Universe was compressed into a tiny particle called a singularity. An explosion tore the singularity apart and turned it into a rapidly expanding fireball.

The early Universe was lit up brightly by the original fireball. As it expanded, it cooled and darkened. (Notice how a red-hot poker darkens once it is removed from a fire.) That is why the night sky is now black. Stars shine because they are still hot. The explosion has become known as the Big Bang.

▲ An incredibly small lump of material, a singularity, explodes forming the Universe.

▶ This cloud, the Veil **Nebula**, is all that is left of a star that exploded 50,000 years ago

EXPANDING UNIVERSE

Within a billion years of the Big Bang, lumps began to form that eventually became the first galaxies. **Astronomers** have found that galaxies are flying away from us. This does not mean that we are at the centre of the Universe. It means the Universe is still expanding.

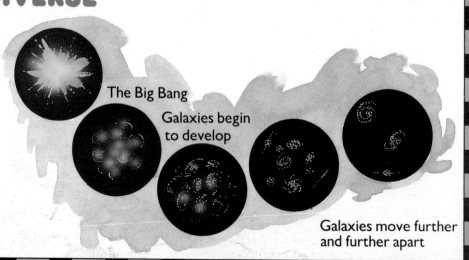

The Big Bang

Galaxies begin to develop

Galaxies move further and further apart

WHAT IS A STAR?

All the stars appear as tiny pin-points of light in the night sky, but they are actually huge balls of burning gas like our own star, the Sun. The Sun looks much bigger than all the other stars, but this is only because it is closer to us than any other star.

A star shines because of a chemical reaction in its centre. Here **hydrogen** is pressed together so tightly by the star's powerful force of **gravity** that its atoms combine, or fuse together. When this happens, they release great amounts of energy in the form of heat and light. This process is called nuclear fusion.

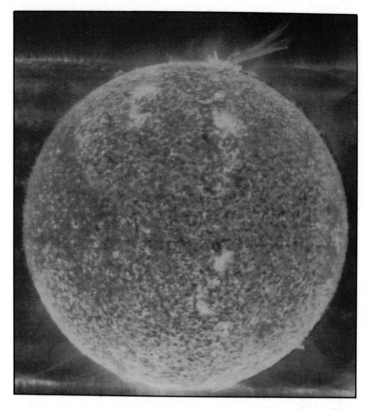

▲ The Sun's surface is like a stormy, fiery sea. Eruptions called solar flares send out radiation that affects radio communications on Earth.

◀ The Sun first formed from a cloud of dust and gas, like the Orion Nebula (right). About 4,500 million years ago, nuclear fusion began and the Sun was born.

Stars are much, much bigger than planets. For example, our Sun is more than a million times bigger than the Earth. Because stars are bigger, they have stronger gravitational forces that can compress and heat the material in their core enough for nuclear fusion to begin. Planets are not big enough to create the right conditions for nuclear fusion. This is why stars shine and planets do not.

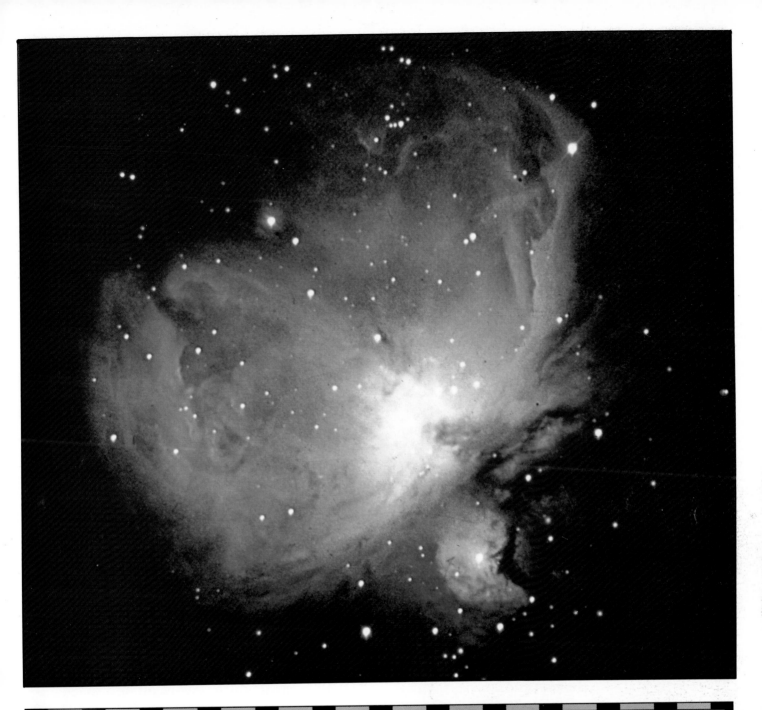

POWER OF A STAR

The centre of a Sun-like star is like a giant hydrogen bomb. Instead of there being a sudden bomb blast, the nuclear reactions in a star keep going because there is a constant supply of hydrogen fuel. The Sun burns 700 million tonnes of hydrogen every second, losing four million tonnes per second in the form of heat and light.

THE DEATH OF STARS

Stars do not exist forever. They are born in vast dark clouds of dust and gas, they shine for as long as their hydrogen fuel lasts and then they die.

In every star there is a balance between the gravitational forces trying to make the star collapse in on itself and the heating effects that are trying to blow it apart. This balance can be kept only as long as the hydrogen fuel lasts. When there is no more fuel, the balance is upset and the star begins to die. Gravity begins to win over the heating effects and the star starts to collapse. However, stars do not all die in the same way. Their future depends on their mass, the amount of matter they contain.

▲ The material thrown off by a 'red giant' is called a planetary nebula. The most famous is the Ring Nebula in the **constellation** of Lyra.

▶ The evidence of exploding stars can be seen in the night sky as rings, shells and clouds of material streaming from dying or dead stars.

A GENTLE END

A small star like the Sun ends its days when its fuel runs out by first puffing up to form an object called a red giant. When our Sun does this, it will engulf Mercury and Venus.

It then blasts its outer layers off into space and collapses to a small object called a white dwarf no bigger than the Earth. Eventually it fades and becomes a 'black dwarf'.

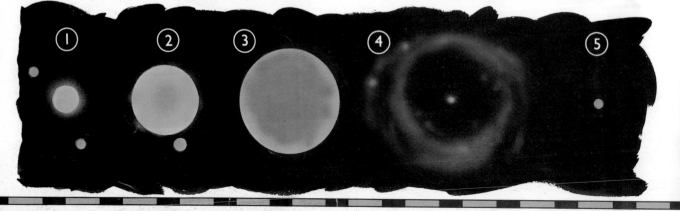

A VIOLENT END

A heavier star swells to form an even bigger red giant when its fuel runs out. Its core collapses but the outer layers explode. The resulting **supernova** shines as brightly as a galaxy. The core often forms a dark, dense object called a **neutron star** or a strange area called a **black hole**, where gravity is so strong that even light cannot escape from it.

BETWEEN THE STARS

The space between the stars is called interstellar space. This space is not as empty as it seems. It contains gas, mainly hydrogen, and dust particles. The gas is so thin in places that there may be only one atom to every cubic centimetre of space – about the same as a drop of water in an ocean!

In other places, the material between the stars is more dense, forming clouds called nebulae. These may be hot and glowing or, if they contain more dust, cool and dark.

▼ The Horsehead Nebula in the constellation of Orion is shaped like a horse's head. It measures three light-years from nose to mane.

As the coolest nebulae do not produce any light of their own, they are normally invisible to observers on Earth. They can sometimes be seen if they are illuminated by nearby stars or if they come between the Earth and a brighter region of space. They can then be seen standing out as black shapes against a brighter background. Nebulae provide the material to make new stars and planets.

By analysing the radiation Earth receives from these nebulae, astronomers have identified dozens of different chemicals, including complicated **molecules** similar to the molecules that led to life on Earth. These life-forming materials may be very common in galaxies.

▼ A black hole contains so much material in such a small space that its force of gravity is huge. It sucks in dust, gas and light.

DID YOU KNOW?

● The Sun sends out streams of particles known as the solar wind. This blows out in all directions from the Sun at great speed, between 250 and 750 km (156 and 470 miles) per second. The solar wind was first detected by the US Mariner 2 spacecraft on its way to Venus in 1982. In theory it should be possible to propel a spacecraft using a huge sail to catch the solar wind!

● There is as much dust and gas floating between the stars of the Milky Way as there is material in the stars themselves. The dense, dusty black clouds where stars form are surrounded by warmer clouds of hydrogen gas.

GALAXIES

Galaxies are enormous rotating systems of stars. They may contain hundreds of thousands of millions of stars, which are all attracted to each other by gravity. Few galaxies exist on their own. Most belong to clusters. Our own galaxy, the Milky Way, belongs to a small cluster – the 30 or so galaxies of the Local Group. Larger clusters can contain up to 2,500 galaxies.

The closest galaxies to us are still so far away that very few individual stars can be seen with ordinary telescopes. To obtain information about the size, age, chemical composition and motion of stars out of view, astronomers analyse other radiations that they send out – radio waves, X-rays and so on.

TYPES OF GALAXY

There are three basic types of galaxies – elliptical (flattened discs), spiral and irregular (no regular shape). Some spiral galaxies have a bar across the middle and these are called barred spirals.

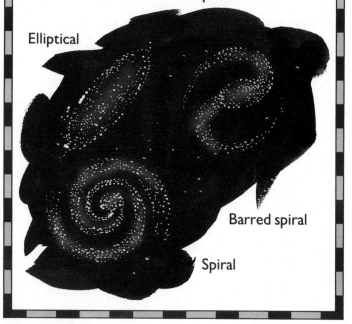

Elliptical

Barred spiral

Spiral

► The Andromeda galaxy is the most distant object that is visible with the unaided eye. It is like the Milky Way, but bigger and with more stars. With a diameter of 150,000 light-years, it is one of the largest spiral galaxies known.

◄ This spiral galaxy is called M51. This means that it is the 51st of 109 bright star objects listed in Messier's Catalogue of Galaxies, Nebulae and Clusters of Stars. The catalogue was created between 1771 and 1784 in France by Charles Messier.

Until the 1920s it was not known whether the swirling star systems observed in the sky were part of the Milky Way or separate from it. The US astronomer Edwin Hubble settled the question in 1924 when he measured how far away these star systems are. Their great distances meant that they could only be separate galaxies. He also measured the distances and speeds of many galaxies.

DID YOU KNOW?

Galaxies rotate about their centre and they also move in relation to each other. The Solar System is orbiting the centre of the Milky Way at about 900,000 kph (550,000 mph). Even at this enormous speed it will take 250 million years for the Solar System to make a single orbit of the centre of the galaxy.

OUR OWN GALAXY

Our Solar System is situated on one of the arms of the Milky Way, which is a spiral galaxy. The galaxy is shaped like a flattened disc with a bulge in the middle. It spins round in space like a giant Catherine Wheel. Scientists have calculated that the galaxy contains 200,000 stars. Our star, the Sun, is 32,000 light-years from its centre.

From another galaxy, the Milky Way probably looks similar to the spiral Andromeda galaxy, although Andromeda is about half as big again.

To us inside the galaxy, the Milky Way looks like a bright band of stars and nebulae lying across the sky, hence the name Milky Way. Many of the stars in the central bulge, or nucleus, are hidden from us by clouds of dust.

▶ This photograph of the Milky Way looking towards the constellation of Sagittarius also shows the track of a **satellite**.

▼ An artist's impression of our galaxy, with the Solar System lying about two-thirds of the way out from the centre.

OUR CORNER OF THE UNIVERSE

The two closest galaxies to the Milky Way are the Large and Small Magellanic Clouds. They are linked to the Milky Way by gravity and streams of hydrogen gas.

There are many galaxies in our corner of the Universe. Our Solar System is just a tiny speck in an unremarkable galaxy in a cluster that is just one of many clusters.

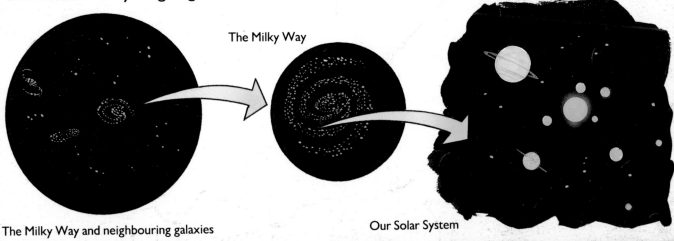

The Milky Way

The Milky Way and neighbouring galaxies

Our Solar System

OUR SOLAR SYSTEM

The Solar System is the Sun and everything in orbit around it. We live on the third planet out from the Sun in a system of nine planets. The orbits of the planets are not circular, but are squashed circles called ellipses. All the planets except Pluto orbit the Sun in the same plane.

Pluto's orbit is tilted to the rest and it actually crosses the orbit of Neptune. Between 1979 and 1999 Neptune is the most distant planet from the Sun. In 1999 Pluto will travel beyond Neptune's orbit and Neptune will become the eighth planet for the following 228 years. Because of this, it is thought that Pluto may not have the same origin as the other planets.

▲ Saturn, known as the ringed planet, is the most spectacular in the Solar System. Its flattened rings, only about 200 m (660 feet) thick, are made from chips of ice and rock.

Each planet moves in two different ways. It spins and it also revolves around the Sun. The time taken for it to spin around once is a day, and the time taken for it to revolve once around the Sun is a year. The tilt of the Earth and its elliptical orbit, repeatedly bringing the planet closer to the Sun and then taking it further away, produces the different seasons.

▲ A band of rocks called asteroids orbits the Sun between Mars and Jupiter. It is material left over from the formation of the Sun and planets.

From time to time, the Earth and Moon line up with the Sun. If one blots out the Sun's light from the other, this is called an **eclipse**. The timing of eclipses can be predicted by studying the orbits of the Earth and Moon.

HOW IT BEGAN

1 Scientists believe that about 4,500 million years ago a star exploded near or inside a cloud of gas and dust. Shock waves made material in some regions of the cloud clump together.

2 The clumps began to collapse under their own gravity, attracting more material towards themselves. The central lump eventually became massive enough for nuclear fusion to begin and it ignited, becoming a star – the Sun.

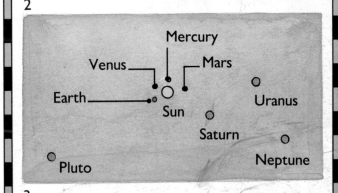

1

2

Mercury
Venus
Mars
Earth
Sun
Uranus
Saturn
Neptune
Pluto

3

3 Smaller lumps of material around the Sun continued to grow by sweeping up dust and gas and by sometimes colliding with each other. After about 10 million years they were the size of the Moon. It took another 100 million years for them to form the planets of our Solar System.

EXPLORING SPACE

The exploration of space began in 1957 when the Soviet Union launched the world's first artificial satellite, Sputnik 1. Since then, hundreds of satellites, mainly US and Soviet, have studied the Earth, the Sun and all of the planets except Pluto. US Apollo astronauts have landed on the Moon.

The most successful unmanned spacecraft were Voyagers 1 and 2, which toured and photographed the outer planets in the 1970s and 1980s. They are now leaving the Solar System to drift amongst the stars.

► The first settlement on another world will probably be on the Moon. The Moon is our closest neighbour, and it could also be used as a base for flights into deep space.

A SPACE MESSAGE

The Pioneer 10 and 11 **space probes** each carried a metal plate engraved with drawings of a man and a woman and details of where the craft came from. Will the messages be answered?

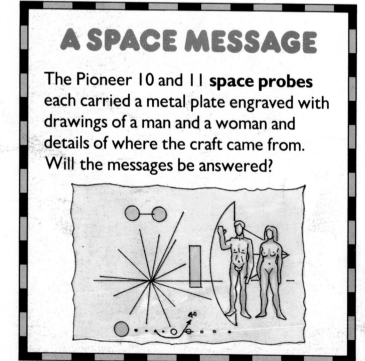

GRAVITY

Gravity is a force that attracts objects to each other. The more massive an object is, the stronger is its force of gravity. The Sun is so massive that its force of gravity is strong enough to hold planets in orbit around it which are up to 6,000 million km (4,000 million miles) away.

Gravity tries to pull everything down to Earth. It is the force that gives us our weight. On a planet with a smaller force of gravity, we would weigh less. On a planet with stronger gravity, we would weigh more. You can do some simple yet interesting experiments with gravity.

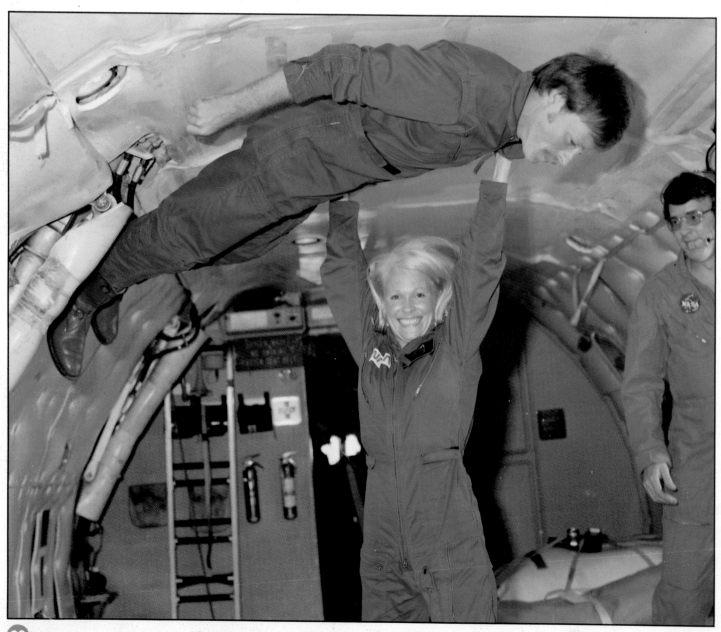

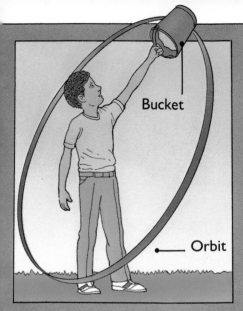

Bucket

Orbit

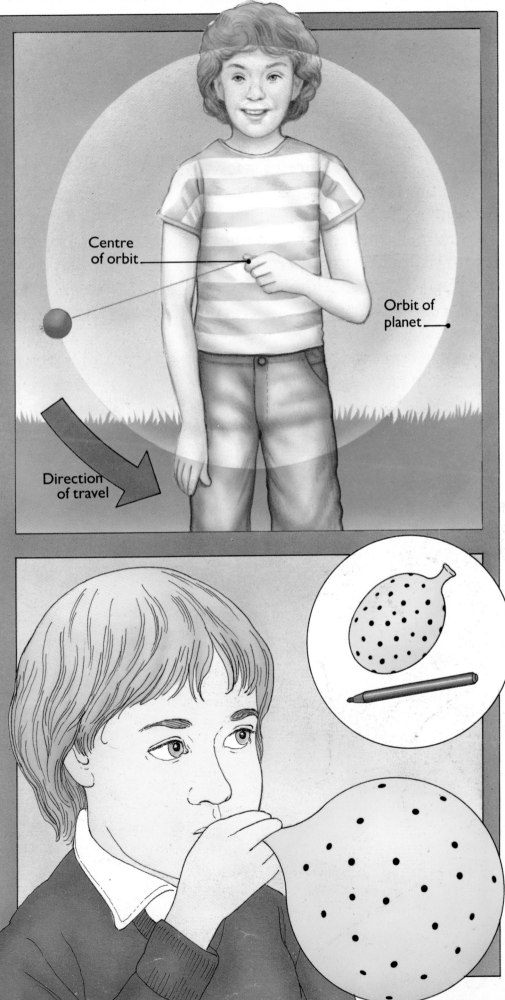

Centre of orbit

Orbit of planet

Direction of travel

▲ Half fill a small bucket with water. In an open place, well away from other people, swing the bucket over your head. The water will not spill out because it will be pushed against the bucket. The bucket is orbiting you like a planet orbits the Sun.

▶ Swing a yo-yo as shown (above right) and let go of it to see what would happen if there was suddenly no gravity. The yo-yo shoots off in a line called a tangent to the circle.

The idea of an expanding Universe can be shown by drawing dots on a balloon to represent galaxies. When the balloon is blown up, the dots move apart.

◀ These astronauts are weightless. They are in an aeroplane flying along a specially shaped course that simulates weightlessness.

Life on Another Planet

This is an imaginary story but it is based on real-life facts and information. The time is early in the 21st century, a year since Julie and her family moved from Earth to the new settlement on Mars. Although things seemed strange at first, Julie quickly settled down and is now enjoying her life on another planet.

Twelve-year old Julie tends the tomato plants in her greenhouse. The tomatoes are ripening nicely. She checks the control panel by the door. A digital display shows what time the lights are switched on, when they are switched off, how much water the plants are given every day and the levels of water and plant food left in the supply tanks. Julie's greenhouse is inside a bio-dome at one end of her family's living quarters. Through the transparent dome, she looks out across an area of red soil and rocks.

The soil is red and the sky is pink, because Julie lives on Mars. Julie and her mother, father and brother live in a settlement of 28 families under a series of protective domes linked by tunnels. Each family has its own private living quarters. The parents work together in a laboratory dome and the children go to school in education domes.

There is a central bio-dome for growing food for the settlement. Each family also has its own bio-dome for food production. This makes it more difficult for a disease or insect pest brought to Mars from Earth to spoil or kill all the settlement's food production. A season's food supplies are also kept in an emergency cold store outside the dome.

The cold store does not need an electrical power supply to keep it going, because the planet's surface itself is cold enough to keep things frozen. From time to time, the food storage tanks are checked. Julie's father, Henry, and another scientist, Rolf, are outside doing that now.

"Rolf, what surface temperature reading do you have?" asks Henry. Rolf holds a plastic box in his gloved hand. A tubular temperature probe extends from one end of the box.

"Minus 43 degrees Celsius," he answers. This is warm for Mars on a midsummer's day. At night in winter, the temperature can reach 150 degrees below zero.

Rolf and Henry are dressed in tough, thick spacesuits. Life support

packs on their backs provide air for them to breathe and also circulate hot water around a layer of fine tubes in their suits to keep them warm or to cool them if they get too hot when they are working hard. The men communicate with each other and the settlement's control room by radio.

Their task finished, they start back to the settlement. Its silver domes sparkle with frost. The domes are silver coloured to reflect as much heat and light as possible back inside to save energy. When Rolf and Henry reach the entrance, the outer door of an air-lock slides open and they walk inside. A flashing sign reads 'DANGER – SURFACE ATMOSPHERE'. This reminds everyone coming in from the surface that they must not remove their helmets yet. There is no air on the Martian surface. The atmosphere is mostly carbon dioxide.

A column of lights in the air lock changes from red to green from the bottom upwards as the surface atmosphere is replaced by air. When all the lights are green, the sign changes to 'AIR: PRESSURISED'.

Rolf and Henry unclip their helmets and breathe in the sweet air as the air lock's inner door swings open. The hissing of their helmet's air supply is replaced by the chattering of people inside and the hum of environmental control pumps and fans.

Henry and Rolf return to their normal jobs. They work together, studying how the raw materials found on Mars can be used to make useful materials for building new structures on the Martian surface or in orbit around Mars. They take samples of soil and rock at different depths in the Martian surface and analyse them to find out exactly what they contain and how they might be processed.

Meanwhile, Julie's brother Curtis is exercising in the gymnasium. Everyone spends some time here. Exercise is particularly important for the children to ensure that they develop properly. The force of gravity on Mars is much weaker than on Earth. Something that weighs 50 kg (110 pounds) on Earth weighs only 19 kg (42 pounds) on Mars. People's muscles do not have to work as hard as they do on Earth and so they do not develop in the same way. Regular exercise helps to offset this.

The family meets for dinner in the evening. Their living quarters are spacious because they cannot go outside for a walk or to play games like people on Earth. Curtis tears open a package that arrived on board the last supply ship from Earth.

"Julie, it's full of video tapes, all the latest movies," he shouts excitedly.

Julie and her mother are watching a screen showing messages from their friends on Earth. They were received by radio earlier in the day and recorded so that they could be played back later. Their recording is interrupted by a noise like a doorbell. *'Ding-dong, ding-dong.'* Julie presses the pause button on the tape player's remote control. An electronic voice says, "Storm warning. Shutters will close in one hour from now. Oxygen conservation measures will operate until further notice."

From time to time, storms blow tonnes of red dust and sand over the settlement. Some of the observation windows and air-lock doors have to be covered by motorised shutters to protect them from pieces of rock thrown up by the strong winds.

Some of the oxygen needed by the settlement is provided by the plants growing in it, but most is made from the Martian atmosphere of carbon dioxide. While the oxygen production plant is not working, oxygen is supplied from emergency tanks and measures are taken to minimise oxygen wastage. Only essential jobs are carried out.

The family decide to have an early night after dinner. As they go to bed, they hear the shutter motors rumble into action. It's the end of another day on Mars.

TRUE OR FALSE?

Which of these facts are true and which ones are false? If you have read this book carefully, you will know the answers.

1. The Earth is one of nine planets orbiting the Sun.

2. The distances between the stars and galaxies are measured in light-months.

5. As all the galaxies we can see are flying away from us, this must mean that the Milky Way is at the centre of the known Universe.

6. The most distant object that can be seen with the unaided eye is a spiral galaxy called the Whirlpool Galaxy.

7. The Milky Way is one of roughly 30 galaxies that form a cluster called the Local Group.

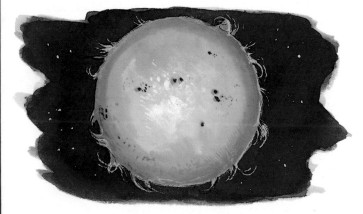

3. The process that makes stars like the Sun shine is called nuclear fusion.

4. The more massive a star is, and many are bigger than the Sun, the more violent the way that it ends its life.

8. Saturn is the most distant planet from the Sun, except for a period of 20 years when its orbit is crossed by the planet Pluto.

9. Saturn's rings are made from particles of ice and rock.

Answers: 1. True; 2. False; 3. True; 4. True; 5. False; 6. False; 7. True; 8. True; 9. True.

29

GLOSSARY

▲ Part of the Veil Nebula, which is the remains of a supernova. The clouds of gas and dust are rapidly expanding.

● **Astronomers** are scientists who study the Sun and planets, the stars and galaxies.

● **Atoms** are the tiny 'building bricks' of all material.

● **Black holes** are thought to be formed when massive stars collapse. No black hole has definitely been identified.

● **Constellation** is a group of stars thought by ancient observers of the skies to look like heroes and animals from their legends.

● **Eclipse** is when one object in the sky passes into the shadow of another.

● **Gravity** is force that attracts objects towards each other.

● **Hydrogen** is the lightest gas of all. About three-quarters of the mass of the Universe is hydrogen.

● **Light-years** are the units used to measure the huge distances between stars and galaxies. A light-year is the distance that light travels in a year.

● **Molecules** are groups of atoms bound together.

● **Moons** are objects that orbit planets. Only Mercury and Venus do not have moons.

● **Nebula** is a cloud of dust and gas in a galaxy.

● **Neutron star** is a small dark object in space caused by the collapse of a dying star. It has an intensely strong force of gravity.

● **Orbit** is the path followed by an object, such as a planet, as it revolves around a larger object, such as a star.

● **Satellite** is an object orbiting a planet. It may be a natural satellite such as the Moon or an artificial satellite.

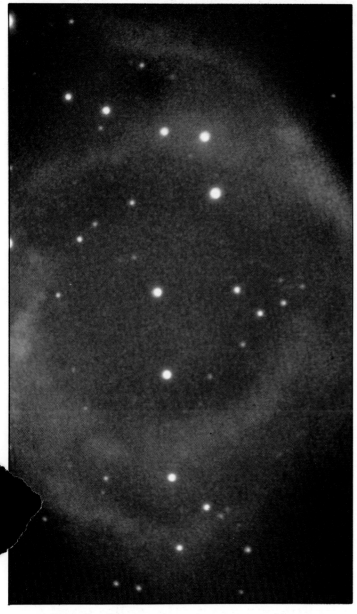

▲ A nebula within the constellation Lyra. It is more than 2,000 light-years from Earth but can be seen with a small telescope.

● **Space probes** are unmanned spacecraft sent out from the Earth into deep space to probe, or study, the planets and the space between them.

● **Supernova** is an exploding star. The star rapidly increases in brightness.

UNIVERSE FACTS

● The Universe may go on expanding forever. If it does, it will gradually cool down and the stars will all go out, leaving a cold, dead Universe. Alternatively, the force of gravity may halt the expansion and the Universe would then begin to contract. It would eventually form another singularity followed by another Big Bang and another Universe.

● Astronomers test their theories about the Universe by studying the stars and galaxies. They obtain information about distant galaxies by looking at not only the visible light they produce, but also at radio waves, X-rays, infra-red and ultra-violet radiations using detectors on Earth and in space. In this way they try to learn more about the Universe, how it was formed, how it works and what will happen to it in the future.

● Nothing in the Universe can travel faster than the speed of light, which is 300,000 km (186,000 miles) a second.

● There is no fixed, still point in the Universe. Everything is moving in relation to something else. We measure the movements of the stars and planets in relation to the Earth, but the Earth moves round the Sun, the Sun orbits the centre of our galaxy and so on. This is the basis of Einstein's Theory of Relativity.

INDEX